SASQUATCH TOOL USE

By David Alan Claerr

Illustrated by David Alan Claerr

Photography by David Alan Claerr
and Michael Brookreson

© 2015 David A. Claerr
All Rights Reserved
No portion of the text, illustrations or photos
may be used without the express permission of the
author, Illustrator or photographers.
The "Fair Use" act does not apply if any monetary profit is made from the use, re-print or any other type of publication of any part of this registered, copyrighted work.

In my previously published research, evidence was presented that our subjects are a sentient species that have many more capabilities than were considered possible in the past decades. Genetic sequencing has determined that they are very close relatives to modern humans and perhaps ancestral in that regard. There is the distinct possibility that, as many Native American sources maintain, they have the genetic capability to crossbreed with modern humans as well. In fact, among many tribes there is the assertion that the Sasquatch are tribes of peoples with whom they often traded or bartered for goods, and shared knowledge of nature lore, medicinal plants and healing techniques.

So, by considering these statements at face value as a viable hypothesis as to the nature of the Sasquatch, a researcher should be able to discover evidence that either supports or negates this hypothesis. In the past decade, my own research has accumulated evidence that the Sasquatch employ the use of many behaviors and technologies that were once regarded as the sole provenance of H. Sapiens or their most recent ancestors, such as H. Neanderthalis H. Hiedlebergensis, and H. Ergaster. Included in these categories are the use of glyphs or symbols that communicate information in the manner of a written language, the use of stones and other materials for toolmaking, and the crafting of items from composite materials for both utilitarian use and as toys for their children.

It is of primary importance to establish these cultural and technological capabilities in the Sasquatch for a full understanding of their nature, as a key step in providing for the protection of what may be considered an indigenous tribal people in the United States and other regions of the North and South American continents.

As a side note, the study of the Sasquatch should be one of unprecedented importance in the fields of Anthropology, Genetic Science and Paleontology. Finding in the study of their biology, social structures and culture may be considered as an analog to a "Holy Grail" in the quest for vital concepts in these fields. They are a living laboratory of an ancient peoples who are even now in a progressive arc of evolutionary development. Their DNA is a genetics "Rosetta Stone" that will be not only a vital key in the understanding of our evolutionary past, but also a prodigious source of information for the understanding of recombinant DNA, vital to genetic therapies for humans.

The bulk of my research in the past decade has taken place in Texas, my home state, and the neighboring states of Oklahoma and Arkansas. All three states have vast tracts of wilderness and areas with exceptional biodiversity and an abundance of plant and animal life that can sustain populations of the omnivorous Sasquatch.

One of the most intriguing aspect of my resent research is the strong evidence that the Sasquatch use not only readily available objects and materials, such as rocks, sticks, bones and thorns as tools, but can also craft fairly sophisticated tools from flint, chert and other volcanic stone. I have also found evidence that they may not only use these tools for processing food, but to also make toys for their children, and to make decorative objects.

One of my first discoveries that indicated tool use was in an area along the Guadalupe River watershed. The Guadalupe and its tributaries run across a large area of Central and South East Texas.

In the area where the first tools were discovered, the dense vegetation along the wide riverbanks resembled a scene from the primordial past. Towering cypress trees formed a high canopy along the water's edges, in some places nearly arching completely over the river itself. Toward the outer sides of the banks was a thick tangle of vegetation. Grape vines as thick as a man's thigh grew in profusion, and many native and imported species of berries sprouted in abundance. Deer were foraging in small groups in every direction, and there were signs of every kind of common mammal in the soil along the banks. In some areas, where there is flat and somewhat sandy, moist soil, such as around the edges of marshes, there can often be found quite extensive thickets of cane.

The cane grows in dense clusters, the 8-to-10 ft. tall reedy stalks growing so close together as to be touching. When I have the opportunity, I will investigate these "cane breaks" as they are sometimes called because I have found that they are frequently used as campsites for the Sasquatch. I look for narrow but well-trodden trails into these thickets. Following the trails, I have found that there will be areas where the cane has been bent and smashed flat to the ground, forming a hollowed out "room" in the cane thicket. The cane forms a fairly dry, clean, mat-like surface that the Sasquatch sleep on and use for their activities. (The first time I discovered one of these encampments, there were at least two Sasquatch sleeping there, during the day, and as I approached, they awoke, and grumbling and complaining with their peculiar, raspy vocalizations, headed out the other direction.)

On another, second occasion when I found stone tools, the Sasquatch campsite was vacant. As I looked about the campsite, I found what appeared to be edged stone blades, resting atop the fresh mat of cane. There was a bit of plant fiber and green chlorophyl staining on the blades. I was not completely sure at the time, but it seemed as though the cane or some other plant material had been cut using the stone blades, fashioned from a type of flint, pictured below. I wondered if they were native American artifacts that had just been picked up by the Sasquatch and put to use. (Other researchers sometime leave gifts of modern steel knives, which the Sasquatch apparently use.)

This area of Texas has some notable sources for this type of chert, which is sometimes call Georgetown Flint.

Tools found along the banks of the Guadalupe River

CENTRAL TEXAS EXPEDITON

In February of 2014, on another expedition in Central Texas within about 30 miles of the site on the Guadalupe River, I found some rather astonishing and significant evidence of tool use among the Sasquatch.

Prior to the expedition, I prepared for researching the area by poring over satellite imagery of the region as well as standard road maps and US Geological survey topographical maps. With the satellite imagery, it is possible to identify objects as small as a meter, in many cases.

By careful analysis of the photographic imagery, I was able to identify the topological features and important elements such as surface water, including springs, creeks and ponds as well as currently dry erosion channels. By using satellite images taken throughout the year, it was possible to identify the dominant type of vegetation, from grassy meadows to cedar groves, stands of cypress and hardwood forests. It was evident that this area of Texas had a unique mix of geologic formations, from buttes and rocky hills to meadow-lands and marshes. The topology is so varied that the terrain presents many micro-climatic ecosystems with abundant and diverse populations of flora and fauna.

I then drove to the destination by car. As I neared the immediate area, I noticed a rather sad but significant fact: on the roadside for several miles approaching the site, there was a high incidence of road-kill. There were snakes and reptiles, small mammals of every common type, and most significantly deer of several sub-species. Some of the deer were imported species and I assumed that they were from the various exotic wildlife ranches and hunting preserves in the area.

On the day that I set off on the expedition, I met with a fellow researcher who had offered to drive to the site. The site was bordered by livestock ranches of several types, and we saw many free-ranging sheep, cattle, goats and llamas. In one area, flocks of sheep and a few cows blocked the road for several minutes, impeding our progress. An exotic wildlife ranch also bordered the site., and as we arrived at the site, there was an impressive amount of deer, antelope, bighorn sheep and other ungulates ranging freely in a sizable meadow.

The investigation site was within an extensive fenced and secured area. Almost immediately after arrival, we began to see unusual signs of activity. When just a few step away from the car, we heard some low grunting sounds. Four small gray deer bolted from the brush, dashing right past us, within a few yards. We went toward the location of the grunts and found a deer skeleton, signs of a kill at that spot some months previous to our arrival.

As we proceeded into the site's interior, walking along a gravely dirt road there were a remarkable number of deer on both sides of the road. A pair of deer, a female and a fawn sprinted from the left side, heading to the road directly in front of us and vaulted the five-foot mesh fence that was topped by a dual strand of barb wire. They crossed the road only about 10 yards in front of us and then vaulted a similar fence on the right side. It was apparent that they were fleeing a threat on the left side of us. The terrain in the immediate vicinity was low, rolling, rocky hills interspersed with cedar and juniper clusters and patches of grassy meadow.

    As we proceeded, we experienced an unusual amount of animal sounds, starting with the alarm calls of birds- jays, crows, ravens and others. As these calls resonated through the hillsides, the air was filled with the most bizarrely loud and cacophonous sounds I have ever heard in all my travels. There were random series of shrieks, howls and hoots somewhat in the manner of primate vocalizations. Then, from all quarters and from various distances away, other animals began to sound off in panic. Horses whinnied and neighed. Cattle bellowed and groaned. In the far distance, bloodhounds bayed and bawled. A flock of turkeys running parallel alongside the road cackled and cawed as they sped past us, dodging through the brush. The effect of all this noise was a bit disconcerting. If I had not known that there were reports of Sasquatch activity in this area, I would have been more alarmed. The fact that few humans walk on foot through these areas, and that the Sasquatch generally hunt and forage at night, may explain the panicked reaction of the animal populations. It appeared that the shrieks and hoots were most likely Sasquatch alerting others of their own kind of our presence, and this set off a chain reaction of the other animals, wild and domestic for a few miles in every direction.

    As we reached the central area of site, along a large creek, there were a series of call-and-response vocalizations. These calls seemed to be an imitation of a Tom-turkeys alarm call, by perhaps three Sasquatch, several hundred yards apart from each other along the creek. The reason that I surmise these calls were Sasquatch imitations is that their mimicked call are often several times louder that the imitated animal's call, and have a noticeable raspiness in the intonation.

    After hearing these types of vocalizations and imitations on several occasions, the raspy tone becomes instantly recognizable, and it is a reliable indicator that the sound is coming from a Sasquatch. Another indication that these calls were not from turkeys was that the intermittent calls drew nearer to our position from different directions on either side, both upstream and downstream along the creek. A real turkey will sound the alarm, then quickly flee away in the opposite direction.

    The call-and-response activity followed us for another quarter of a mile or so. The area was teeming with life. In the creek there were several species of fish; among them were bass, bluegill and various minnows. There were large freshwater prawns and crayfish- evident from the remains of their exoskeletons on the banks. Towering cypress trees lined the creek. There were even frogs that had emerged early in the year, lured out by the unseasonably warm, balmy weather. In fact, the weather conditions that day were exceptionally fine for Texas at any season: sunny with temperatures in the high 70's (F) and a light breeze steady from the south-east.

We then proceeded to follow along a dry tributary creek bed that had carved a narrow ravine through the valley, running up-slope into groves of cedar. From the satellite imagery , I had determined that this creek-bed led to a spring. Sasquatch need to drink probably a few gallons of water a day, and like humans, they prefer the clear, fresh water of springs. So areas around spring-heads often show tracks and other evidence of their presence and activities.

The dry creek bed showed evidence of the harsh changes in the environment due to the previous few years of extreme drought. For example, archived satellite images going back several decades show that the spring flowed all year long and that the spring was a perennial source of fresh water. This year, not only was the creek dry, but the water table (underground) had dropped so low that there were sinkholes in the creek bed The sinkholes were formed when the water-flow of the spring, normally above-ground, flowed at a reduced rate a few yards below the usual creek bed, undercutting the topsoil and overburden of sediment. In some spots, the underground channels had collapsed, and sinkholes were formed. Runoff from rainwater then seeps into the sinkholes, enlarging them.

Some of the fences along the property lines crossed this tributary creek. The fences were of different types, from several eras, dating from perhaps the early 1900's to the present day, judging from the amount of corrosion, differing styles of barb-wire and mesh patterns. All of these fences, and in fact, all the fences along the perimeters of the extensive private property had been compromised, in that there were breaches created either by bending and forcing-up sections of the mesh fences from the bottom or, in some cases, with holes created by a repeated bending of the wires until they became brittle enough to break. Well-worn paths often led through the fence-breaches, and there were prints from deer, javelina, feral hogs, raccoon and possum in the dusty areas, as well as the large, broad, though indistinct tracks that are often left by the Sasquatch.

Steel fence-wire, bent repeatedly until broken.

In proceeding up the creek bed, as the land rose into a more level plain, and we began to see evidence of an unusual type of activity often attributed to Sasquatch. The evidence consisted of cedar branches that are twisted around several times. Some are twisted until they have broken off. The twists are usually in a clockwise direction, as though by a primate with opposable thumbs, and using the right hand, as do the majority of humans. To twist off a branch that is perhaps two inches in diameter takes a considerable amount of strength and perseverance. I tried doing this myself- it took the use of both hands and five to seven twists of a full 180 degrees rotation to accomplish this. And, as they are twisted, the fibrous branches become tighter and harder to twist. I was wearing leather work gloves as well, and without them, I would not have been able to break the limb by twisting. In the area throughout the day we found about thirty or more branches that had various degrees of twisting. Another feature of the twists is that there is no evidence of a feeding activity- the cedar needles and bark remain on the branches, intact except for bruising and tears. So why would Sasquatch engage in this activity? A viable hypothesis is that it is an instinctive behavior that enables the young, from toddlers to juveniles and sub-adults to build up the strength and coordination for the grasping and wrenching motions that are needed to capture, kill and ultimately process the larger prey, such as deer and other ungulates that are part of their omnivorous diet. And indeed, later in the day we found evidence of this hunting activity, as you will see in photographic evidence to follow in this book. In a few areas, there were numerous twists, as though several young Sasquatch may have been competing in tests of strength. Some of the branches were small and low to the ground, evidence that even young toddlers begin to engage in the activity.

**Cedar branch twisted twice around-
a full 720 degrees of rotation.**

The cedar groves leading to the spring-head were in a fairly level area in a central valley. When we reached the spring-head, it was not flowing above the surface, though the soil was damp and water could be reached by digging down only ten inches or so. There was little evidence that the spring was currently in use then, although there were many indications that it had been extensively visited in the past. Since the flow had been interrupted by the drought, it is more likely that the still-flowing main creek was being used as a water source by all the native wildlife as well as the Sasquatch. But in the general area of the spring-head, perhaps 50 yards or so to the east, there was an area with very unusual signs of activity. There, grassy patches of a low-lying moist meadow had been extensively torn and turned over. At first it had the appearance of a rooting activity by feral hogs but, except around the edges, there were no hog footprints, and the total area where the soil had been overturned was much larger than any I had previously seen that was created by hog-rooting.

When searching around the perimeters of this area, I found something remarkable- the large, long antler of an Axis deer- a species imported to the area, native to the sub-continent of India. The remarkable aspect was that the antler showed signs that it had been used as a digging and rooting tool in the grassy areas. There was dirt stuck to the antler points. and grass-stains up to five or six inches above the point tips. Areas along the shanks of the antler showed wear from being held in the grasp of a prehensile hand. The logical assumption is that the Sasquatch were using the antler as a tool to dig up a sub-soil food source such as worms, grubs or edible fungi. The most likely explanation is that the Sasquatch were harvesting Yampa, an edible tuber that was a staple food source for many Native American tribes west of the Mississippi. (There is a tribe known as the Yampa, due to their reliance on this crop. The name Yampa was considered for the State of Colorado in the early years of its settlement.) The antler was the first in a series of objects and artifacts that very strongly indicate tool use by the Sasquatch.

The Axis Deer antler and a demonstration of its apparent use, indicated by soil stains on the points

We continued our investigation of the area by first circling the perimeters and then working our way into the interior. Previously, in viewing the satellite imagery, I had noticed features on the landscape that had several of the earmarks of an ancient, Pre-Columbian stone-building culture. There were some very large stone blocks, some still placed atop others, in several locations, all near water sources. In other areas, there were linear structures that appeared to be remnants of stone walls or fortifications. I had determined that one such wall was slightly to the southeast of the spring-head, on the border of the perimeter, so we worked our way in that direction. We first found what appeared to be the ruins of a village that was comprised of circular stone-walled huts. The remnants of the stone walls were scattered about in roughly circular patterns, and the stone blocks had evidently not been secured by mortar. The stones of this village site did not show the pronounced weathering and discoloration of some of the more ancient blocks that we later found, so by comparison they were probably some few hundreds of years old. The wall, which was nearby, showed a discoloration of the limestone (a deep gray tone), considerably eroded surfaces, with growths of lichen, and had the appearance of cut stone that was perhaps a thousand years or more in age. The wall was partially buried, and over 150 feet of its length was projecting above-ground. It is likely that the wall extended 10 feet or more below the ground, since it rose 4 feet above ground at the highest. The top of the wall was fairly level and straight, and the blocks were hewn and fitted with a fair amount of precision, though there was no mortar to secure them. These ancient ruins are very intriguing and perhaps merit further investigation by an archeological team.

Section of Pre-Columbian stone wall

In continuing with the investigation, we proceeded across some pasture land that had been cleared and perhaps cultivated by modern humans at some time, but that now lay fallow. At the edge of this pasture land was another spring-head, larger than the first, and surrounded by mesh-and-barbwire fencing. The fence had a livestock gate, and the small pond surrounding the spring had been used by the farm and ranch animals. The pond was mostly dry with a patch of damp mud at the center. On one side of the enclosure, the mesh of the fence had been forcibly pulled up from the bottom edge, and bent back to form a gaping hole that a human, or something considerably larger could fit through. From this gap, a set of bipedal prints lead straight across the pond to the other side, where the water in wetter times would flow down a steep cascade of boulders. The tracks were perhaps a week old and somewhat indistinct. We photographed the tracks, after close examination, and continued down the rock cascade.

Bipedal Tracks in the mud of a dry spring-fed pond

The now-dry cascade of rocks was impressive, and showed that in the recent past, this would have been a picturesque setting with a waterfall tumbling down about 70 feet and dashing against the well-worn boulders, many the size of automobiles and pickup trucks. As we descended down the cascade, amongst the jumble of boulders, we found a few deep recesses, one of which had the tracks and signs that it was the den of a cougar. And, as we clambered down the boulders, there was movement in the surrounding brush and woods- something large- but we could neither see it nor determine what is was.

At the bottom of the cascade, the watershed formed another dry creek bed that ran along a well-eroded ravine. The deep and broad ravine showed evidence that the spring had produced flowing water for thousands of years on a regular basis, prior to the aquifer in that region having been tapped and severely depleted by humans for agricultural and urban use. The bed of the creek was cut down right to the limestone bedrock and the bedrock itself had an eroded channel worn by the creek where it flowed.

As we neared the area central to the reported Sasquatch sightings, we began to find evidence of activity. Along the banks of the creek there were scattered horns, antlers, bones and skeletal remains of deer, sheep, cattle and other, less easily identifiable animals. There were also a few sacks of animal feed that had been pilfered from a livestock barn- they were ripped open (not cut or opened by pulling out the drawstring). These signs made me feel confident that this was an area frequented by the omnivorous Sasquatch.

As we neared the ranch house and barn inhabited by our hosts, who had reported the activity and invited us to investigate, the creek bed leveled out to a wide and flat surface of limestone- nearly as featureless as a cement sidewalk, but covered with small scree comprised of limestone fragments.. But right in the middle of this bare tract, there was a pile of boulders placed right in the middle of the dry creek. Some of the boulders had toppled from the stack. This placement of the boulders is a familiar device often referred to as a cairn - a marker that is used by the Sasquatch to denote important locations. (Yes, humans also use these markers, but the largest of these boulders were of such size and weight that they could not be moved and stacked by less than perhaps five or six humans without large machinery, and the depth and narrowness of the ravine would not allow such machines to enter.)

**Boulder Stack, or Cairn.
The scale is suggested by the leg of the adult male at left.
The topmost rocks have tumbled off and fallen forward.**

The discovery of the boulder cairn set off an immediate reaction- I had a sudden rush of adrenaline- complete with the symptoms of goose-bumps and hair raising up on the back of my neck. I had the conviction that this was a very significant area. I began to scan the bank of the creek for a land feature that I had seen on previous expeditions- a small level area or plateau that would be use as an encampment or periodic habitation. And there, perhaps 20 feet above the boulder cairn was just such a level, grassy shelf, shaded by towering oaks and elms. There was a worn pathway up to the small plateau. There was more evidence of habitation by the Sasquatch- bones, antlers, feed sacks, etc.

It is important to note that this Sasquatch campsite, probably used periodically over centuries, was directly below one of the ranch's newer outbuildings- a guest house built above the creek, near the edge of the ravine. This is significant because many of the artifacts and items that we discovered in and around the Sasquatch campsite had been scavenged from the ranch. Besides some of the manufactured items, which will be detailed later, there were bits of food-related items- grapefruit rinds, oranges (probably tossed out whole when a spot of mold was seen) that were bitten open, with the juice sucked from the pulp. There were margarine tubs and tin cans that appeared to have been licked clean. There were table-scraps such as t-bones from steaks, spare-rib bones, ham bones and pork-chop bones that had been nibbled on and stripped clean. These items were very obvious when we first arrived at the campsite.

There were also skeletal remains of many ungulates. I noticed that several of the leg-bones had been split open. Close examination revealed the impact of a pointed instrument indicated by chips and bone fractures where the point struck the bone. The object of splitting bones is to remove the nutritious marrow. I started to methodically search the campsite working across from edge to edge in a grid pattern about a yard wide per grid-square.

It was then that I made what I believe is a momentous discovery- I found a hand-ax crafted out of chert. The hand-ax closely resembles paleolithic hand-axes crafted by prehistoric humans and their ancestors. The point of the hand ax matched the impressions in the bones.

The Paleolithic-style Hand - Ax.
Made of Chert, often referred to as Georgetown Flint

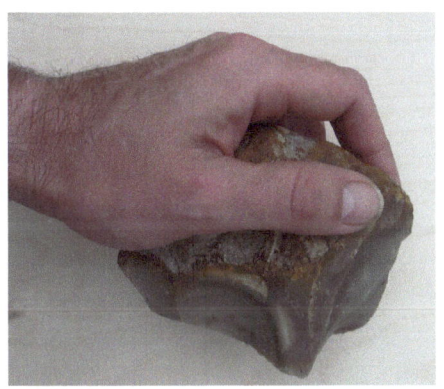
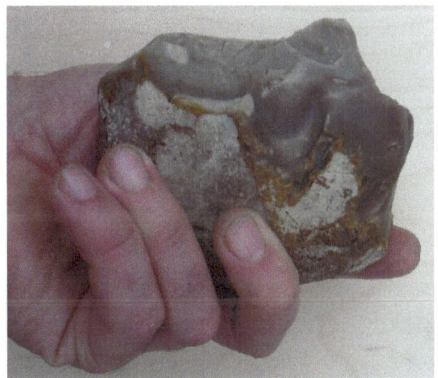

Point of the Hand Ax

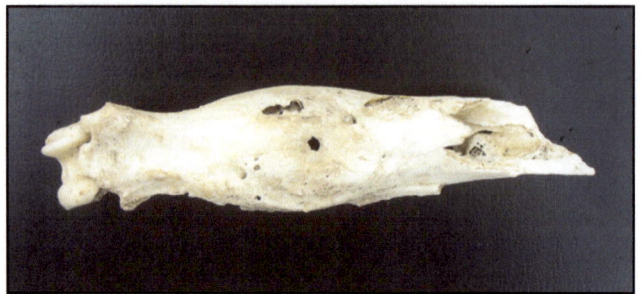

One of many leg-bones split with the Ax.
Interesting because it shows a broken limb that had healed.
The animal would have been lame and thus, easy prey.

There are several facts about the hand ax that are pertinent. The first is that it was found lying on a patch of soft, fine, green grass. The grass beneath the hand ax was fresh and green, which indicated a very recent use and placement. The ax itself had many features that showed that it was crafted very recently and that it was not a artifact from the distant past. The ax was free of oxidation and iron-staining, which are present in stone artifacts that have been buried. There were no indications of erosion on the surface. On a freshly crafted flint or chert tool, there are fractures along the edges of the chipped and flaked-off areas. The fractures on the hand-ax were free of stains from soil, and were nearly white, as they are in a newly crafted stone tools. By contrast, in tools that are of ancient origin, the fractures will have impacted soil deposits and be discolored by soil or water-borne iron. The cutting edges of the hand ax were sharp and showed little if any signs of wear except for the striking point, which evidenced multiple impacts on a hard material, such as bone.

In the same encampment area, I also found a flint core- which is a nodule of flint from which large flakes are struck off, using a hammer-stone. The large flakes of flint have sharp edges, and can be used as cutting instruments. Alternately, the flakes can be knapped, or, further worked into knife blades, arrowheads, spear-points, etc. There was also a large flake from another core, not the one found in the encampment.

Flint core and a cutting blade, flaked-off from a similar core of a different material.

Another very significant find was a tool used to "knap" or flake off small chips from the cutting edge of a stone blade, ax or other implements . The tool was a piece of deer antler that was sharpened to a point. The point of the antler knapping-tool is pressed firmly along the edge of the implement to create a sharp serrated edge on the implement. Finding the antler knapping-tool with the ax and the core is very strong evidence that the toolmaking took place in very recent times- only weeks or months prior to their discovery.

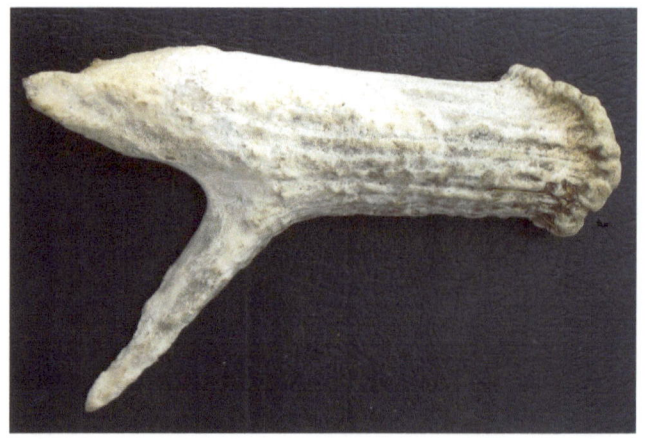

Flint knapping-tool fashioned of deer antler.

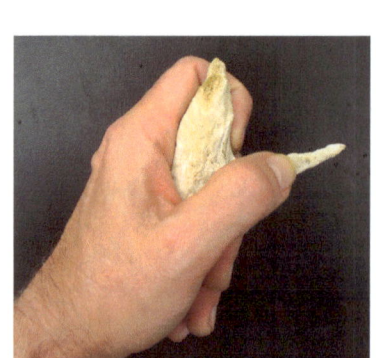

Hand-ax shows marks from this very tool,
used as demonstrated above.

Sasquatch splitting a bone to extract the marrow

In carefully searching the campsite and immediate area, we found a considerable amount of somewhat surprising evidence of sophisticated behaviors that are indications of the intelligent, creative and curious nature of the Sasquatch. Keep in mind that we are discussing families of sentient beings, which could possibly be human ancestors or a closely related species. They have among them babies, boys, girls, mothers, grandmothers and grandfathers that outnumber the large, mature males of popular imagination. Sasquatch habitually avoid conflict with humans and, when left unprovoked, are one of the most peaceable species on the planet.

Since this expedition took place in February, the tree and brush were clear of leaves, and the weedy ground cover that is thick in summer, was sparse and dry. So the search of the campsite was made much easier than it would have been otherwise. In previous explorations, I had found that Sasquatch will often have spots where they will place collections of items that are non-utilitarian, and that have value as toys, objects of art, or as curiosities and perhaps talismanic symbolism. At the west edge of the campsite was one such spot, underneath a dense bush, whose overhanging branches would have completely covered the collection most of the year. The collection consisted of items both natural and man made, and the items had elements that, in human society, are often correlated with feminine interests and proclivities. For example, there was a piece of pottery shard that had a colorful, decorative flower motif. There were items of differing materials that were grouped according to color, such as an orange plastic cap placed together with orange weed-cutter cord and bits of orange plastic from a broken toy. There was an unusual fossil clam-like bivalve, with both halves of the shell cemented together, forming a symmetrical object of beauty.

Clam-like Fossil

Various items from the collections at the camp

In another spot, farther to the east and uphill a bit, also at the base of a thick bush, there were other objects in a collection that had attributes associated often with a masculine nature. Parts of tools were cached there, with lengths of sturdy rope. There were a few target arrows from a child's archery set, together with only one half of the fiberglass bow. The pieces from this archery set were weathered and rusted, and appeared to be perhaps a decade old. But a very significant find was what may very well have been a toy crafted by a Sasquatch: a bow. The bow is particularly interesting because it is not functional; rather it is an imitation of a bow, made from a naturally curved branch of dry oak wood, stiff and unbendable. The bowstring is made from a length of the thick rope, evidently cut with a rough tool (most likely the hand ax) and tied to the bow with clumsy knots.

Toy Bow fashioned, perhaps, by Sasquatch, at bottom. At top is: half of a child's fiberglass bow and two old aluminum target arrows. The toy bow is a rigid, naturally curved branch that does not bend.

Picture this scenario: A juvenile Sasquatch, hiding in dense cover, watches a human child shooting the arrows from his bow and is intensely curious about these objects and activities. Some years later, after the child had perhaps moved away and left behind the now broken bow and a few tarnished arrows, the Sasquatch finds the pieces of the set, and tries to craft a bow. Although he or she knows the shape of a strung bow, from watching the human child use them, the concept of using a springy piece of wood to fashion a working bow is unknown. So, like many human children, the juvenile Sasquatch just plays, and pretends to shoot the arrows from the non-functional toy bow.

Other items both in these collections, and scattered about the camp, were: empty glass jars, light-bulbs, pieces of toys and bits of old electronic devices, such as radios or phones.

We proceeded up the high creek-banks toward the ranch guesthouse above the camp. Behind the guesthouse was a small fenced yard. The gate in back had a latch, but no lock. There was a curious bit of evidence of possible Sasquatch activity there, a feature that is perhaps an indicator of their intelligence. A garden hose was attached to a free-standing spigot in the yard, near the house. The hose had been stretched straight along its full length, strung across the yard. The hose passed under the back gate, where it extended a few feet, almost to the edge of the sloping creek-bank. It appeared that the Sasquatch knew how to turn the water on and off at the spigot, and had arranged the hose in that position for their own use, drinking the water. Later, when questioning the residents of the ranch whether or not they had placed the hose in this position themselves, they denied having done so and, checking the hose themselves, were baffled as to how this had occurred. "Who would run water down to the creek?", they asked; "Tap water out here is too precious to waste like that."

In searching the yard, we also found that a padlocked side gate that led to a small vegetable garden had been broken open with such force that the hinges of the gate were torn off the gatepost. This was another occurrence that baffled the ranch residents, since it would take strength beyond that of an average human to break down a sturdy wood-frame gate like that.

On the east side of the house, opposite the gate side, we found other possible evidence of Sasquatch activity. In a small naturally eroded ditch or runoff leading down to the creek, there were scatterings of the remains of table scraps- fruit and melon rinds, bones from processed livestock, cans and plastic containers. It is significant that in the runoff ditch I also found a small flint blade of recent manufacture. But there was also what appeared to be a symbol that was made of tree branches that had been placed in a pattern somewhat in the shape of a capitol "A", similar to others that both my research companion and I had seen in other areas with Sasquatch activity. Symbols such as these are often directional markers- in this case perhaps indicating the location of the campsite in the plateau below. [The reader may refer to my previous book Sasquatch Genesis (available as an e-book through Amazon.com), which contains photos of symbols and an essay discussing their use.]

Symbol or Marker in form of an "A"

Proceeding with our investigation, we headed east to the area where the main ranch house, barn and corrals were located. Horses and a few cattle were kept in a pasture-field nearby. We searched the area thoroughly but there were no overt signs of activity until we came to the front yard of the main house.

In the yard was a unique and somewhat peculiar decoration constructed by the ranch owners- a "Christmas tree" that consisted of a welded metal framework on which were hung deer antlers and a few of the horns of bighorn rams. Most of the deer antlers had been drilled-through and had wire hooks attached. There were many antlers, though, that had no hooks. None of the ram-horns had hooks or holes. In the yard around the antler-tree, there was a scattering of skulls, bones and antlers. The bones were mainly leg bones. There were a few deer-skulls, and a skull from an exotic antelope, complete with horns. It almost seemed that the bones had been arranged in a pattern at one time. These patterns, though a bit indistinct, intrigued my research companion. He had seen previous examples of both antler stacks and patterns of bone and antlers at another site, where he had conducted long-term investigations.

At this other site, he and other researchers had interactions with the Sasquatch that involved the periodic placement and rearrangement of bones and antlers, along with other objects. Both the researchers and the Sasquatch would alternately place and/or rearrange objects in patterns- the researchers by day, and the Sasquatch by night. At the other site there had been other interactions as well, such as the setting out of food and other gifts that were accepted by the Sasquatch. Often gifts were left in reciprocation by the Sasquatch, such as colorful feathers, pebbles and other interesting objects.

The antler-tree, with antelope skull we had later placed on top.

The immediate vicinity of the ranch house had several remarkable aspects. The house was situated adjacent to the main creek, in which water was still flowing year-round, although at a rate significantly reduced following the years of drought. This creek is a sizable tributary to one of the main rivers in the region. Along the creek in this area, there are stands of magnificent ancient cypress trees, many of which, judging from their enormous girth are from 500 to 800 years old and some of the true giants, even older.

An immense Cypress Tree.

The person is 6'2" tall. Identities of the persons in this collaborative study have been kept confidential, and specific locations are not revealed due to the attempts to hunt and kill Sasquatch, by others.

Near one of the largest of such cypress trees, but up on level ground above the banks we found what for all appearances were burial mounds with much the same size and configuration as Native American burials. We noted the mounds, but left them undisturbed. (There are in fact numerous archaeologically certified burial sites in the area for several miles around.)

Possible Burial Mound

But along the creek upstream perhaps only twenty yards from the giant cypress, there was an artifact that had the appearance of a rudimentary ambush-blind. Interestingly, it was constructed with a section of mesh-fencing about four feet by eight feet that had been bent in half to form a roof for a small tent-like hut.

[Later we found, about a mile and a half further upstream, the place from which the section of fence had been removed. The 4'x8' piece had been part of an extension of the fence that crossed the creek. The piece had originally been placed below the fence-line to extend down into the creek, preventing game and humans from going under the fence via the creek.]

The hut was propped up by thick sections of branches- broken, not cut- and the roof of mesh was completely covered in a thick mat of cypress needles. Other small branches that were broken-off from bushes along the banks, were piled around the hut for further camouflage. One side of the blind was left open as a door. It seems evident that this hut, which was a bit too small for an adult human, could have been made and used by juvenile Sasquatch as a hunting blind- serving for a well-concealed ambush.

Ambush-blind constructed of mesh-fencing with rough branches for support. Covered with a mat of cypress needles.

Ambush-blind viewed from front opening. The tent-like structure opening at front faces out to the creek.

Our investigation had begun in the early-morning hours and continued through into the late afternoon without a break. At that time we then met with one of the residents of the ranch and had the opportunity to discuss the reports of the activity and ask questions about the situation there. Along with the typical reports of hearing unusual primate-like sounds- hoots, howls and the occasional scream, we also asked about the antler rack and the bones and skulls around it. The reply was that the bones on the ground and the antelope skull were not there as part of the original decoration- the antler tree had only deer antlers, and was usually strung with Christmas lights in December. The resident said that the skulls and bones, as well as additional antlers, were bought in periodically, mysteriously appearing overnight. And then, the ranch resident reported that "Every night the raccoons come and take all the antlers off the tree, and I have the caretaker put them back on the tree every morning." We asked if she had ever seen "the raccoons" remove the antlers. She replied no, she had not, it always happened late at night.

We then realized that this was a prime opportunity for an experiment. Since we planned to return the next day, we then placed all of the antlers on the tree, and topped the tree with the horned antelope skull. The top of the tree was about eight feet high, and this would have been very difficult if not impossible for a raccoon to reach, having to first climb over the tangle of antlers, wire and electrical chords. The skull was firmly set on the pointed metal spike that formed the tip of the tree, and it would have been also very difficult if not impossible for a raccoon to dislodge it.

The next morning, the skull had been removed from the tree and placed at its base. Most of the antlers were undisturbed. Some of the antlers, however, were removed from the tree and arranged to form a rectangular border on the north side of the tree. Now, given the fact that there were only two people in residence at the ranch that night- one was the caretaker, a hardworking Hispanic gentleman who habitually rose at dawn and turned in at sunset, and our host who had spent the day in a distant town overseeing a business venture, it is unlikely that we were being hoaxed by either one of these people who clearly did not have the energy for such nonsensical mischief in the middle of the night.

The entrances to this area have a series of locked and monitored gates, so it is also unlikely that anyone from outside the ranch was entering at night to move the skull and antlers.

Antler tree the morning after the experiment. The Skull at top was removed, and several antlers were placed in a rectangular pattern aligned to true north

After taking photographs and videos of the changes to the antler tree and surrounding vicinity, we met with another of the area residents that had reported the activity, had experienced visual sightings of the Sasquatch, and who had also photographed Sasquatch previously. A group of us then departed on a survey of the eastern-most area of the ranch. Riding in an ATV and in an auto, we passed through meadow-lands interspersed with cedar groves, to an extensive area where high limestone buttes rose out of the plains. We stopped in several locations to look for signs of activity. Throughout the meadow-lands, there were several herds of deer of various species grazing on the tall grass. In some areas, we found the familiar cedar twists and also gatherings of rocks in piles, and some of these piles contained the smooth, gracefully shaped fossil shells. In these locations we also found skeletal remains of several animals.

In passing by one meadow, we spotted the remnants of a fresh kill lying in a patch of sparse grass on rocky soil. We halted and inspected the remains, which consisted of a single leg from one of the huge imported Axis deer. The Axis deer often have a spotted or mottled coat.

The leg was from the hindquarters, and included parts of the haunch together with the entire fore-leg and split hoof. But there were several notable elements indicating that the leg was part of a Sasquatch kill. The first was that the haunch bone had been broken off from the rest of the carcass with a single, sudden twist-and-snap motion. This was evident from the clean, diagonal fracture of the bone. The second indicator was that the skin from the entire haunch was peeled off and pulled back over the rest of the leg, in the same manner that one removes a glove from one's hand. Although there were other predators in the area; (later I saw a cougar sitting sphinx-like on a low mound on a small prairie-like plain), big cats, wolves and other carnivores typically either chew and/or tear the skin off their prey. Only a creature with strong fingers and opposable thumbs could so deftly peel back the thick hide and double it back over the leg. The third indicator was that the muscle tissue appeared to have been pulled off the bone, rather than bitten off, and there were none of the signature tooth-marks usually left by a four-legged carnivore.

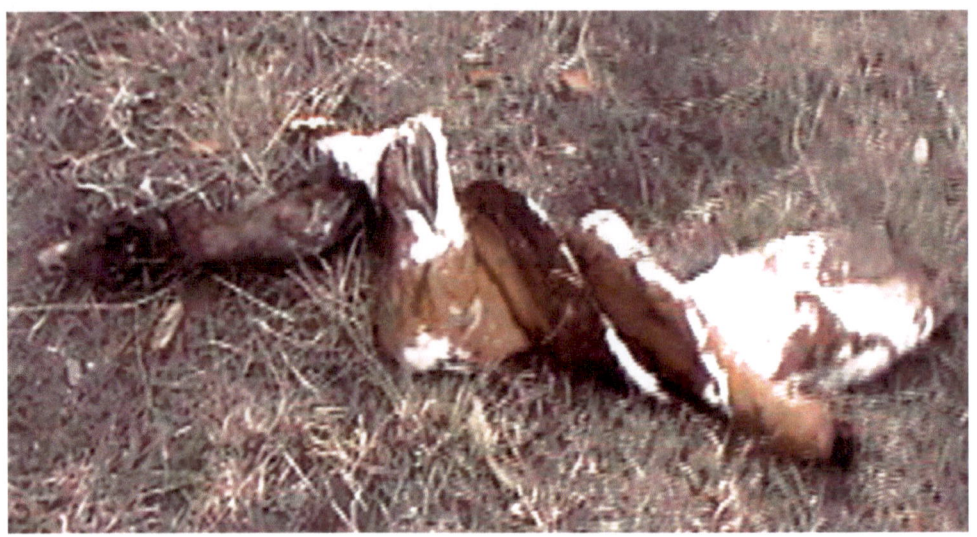

**Axis Deer Leg-
The hoof is at right. Pelt has been peeled back like a glove.**

We continued the tour of the ranch. We had the opportunity to scale the highest butte on the ranch, which towered hundreds of feet above the surrounding countryside. The view from the peak was magnificent. The ability to look down with binoculars into the valleys, ravines and creek-beds below offered a greater understanding and appreciation of the highly varied terrain, replete with dozens of pocket ecosystems that harbored an astonishing diversity of flora and fauna.

Prior to leaving on the tour, our host, who had, in other locations, left gifts of food for the Sasquatch, that day had set out tangerines, placed at the base of the giant cypress tree along the creek near the ranch house. When we checked on the tangerines only about two hours later, all of the dozen-or-so tangerines had been bitten into, and the juice had been sucked from the pulp.

There was a line of tracks that came down the embankment leading to the cypress tree. The impressions left in the thick mat of cypress needles were from a bipedal hominin. The length of the prints were about eight-to ten inches, and they were impressed deeply into the needles and topsoil below, indicating that a relatively small juvenile Sasquatch, somewhat heavier than the average human, had dashed down the embankment and sampled the tangerines, then made a hasty retreat. Given the time frame, this probably happened while we were still photographing the re-arranged antler-tree earlier that morning.

I followed the tracks back up-slope. They passed by the mounds we saw earlier that resembled grave-sites. Where the tracks crossed a small but steep ravine, there was a distinct, rounded heel imprint in a bare patch of dirt, together with less distinct prints leading down the steep slope there. (On a subsequent expedition, we found a bedding mat of plucked grass on a ledge in that ravine that was completely enclosed by grapevines and brush. The bed was near where we saw the heel print.)

After returning to the ranch house, we discussed our findings and made plans to set up video surveillance focused on the antler tree. My research partner also re-arranged the collections of artifacts and items that we had found earlier at the campsite, with the intent that they would be monitored for interactive re-arrangement by the Sasquatch, as often occurs in situations where Sasquatch become habituated to humans.

Although this expedition was one of brief duration, we regarded it as a preliminary to a long-term study that would perhaps involve a degree of habituation and interaction with the Sasquatch. Our first priority then is to document their presence, primarily with clear and definitive video sequences, and secondarily with high resolution photographs. To continue with the study, it will be necessary to have financial backers that will help with the purchase of the necessary equipment and to cover the expenses of more lengthy investigation and monitoring.

I was also given the opportunity to examine some specimens of hair collected by our host at another site, only twenty-odd miles distant. Under microscopic examination at 900x magnification, the hairs evidenced characteristics of Sasquatch hair. The surface of the hair had an "imbricate" scale pattern, which Sasquatch have in common with humans and other primates. With adequate back-illumination, it is also possible to see into the interiors of the translucent hair. There was another evident Sasquatch characteristic, a medullar core that was manifest in randomly intermittent segments. This pattern is unlike that of humans and most mammals, which have a continuous medulla that runs through the center of the hair for most of its length.

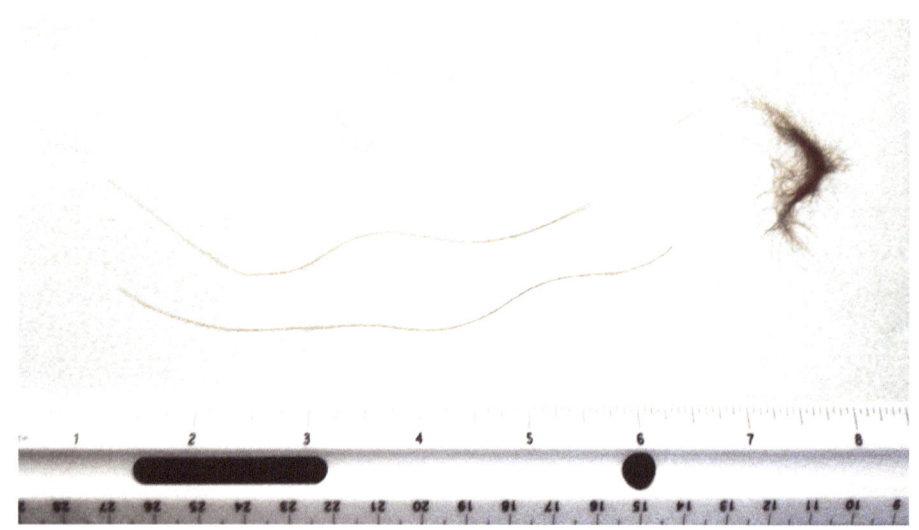

Presumed Sasquatch Hair
As of this writing, the hair has been sent to a laboratory that specializes in identifying primate hair.

    Our host also had previously taken several photographs at another location about 13 miles away. The low-res photos were taken with a mobile phone.
    The best of these images is presented below. In the photo, a juvenile Sasquatch, most likely a female, is creeping on her belly through a thick brush-pile, toward a blue plastic bucket that had watermelon and other fruit placed in it. Remarkably, she is clutching a pink plastic baby-doll in her right hand. In her left hand is another toy which appears to be a stuffed teddy-bear. She is pressing the teddy-bear against her left cheek, obscuring her face slightly. Our host believes these toys were taken from their storage shed when they were residing on the property previously. The figure in the photo has only been color-enhanced to show the brown-coated Sasquatch Girl more plainly. In the front right side of the photo is a broken-open watermelon, which was also color enhanced for clarity.

This endearing photo shows the very human-like affectionate and empathetic nature of most Sasquatch. Would not hunting and killing them be akin to murder?

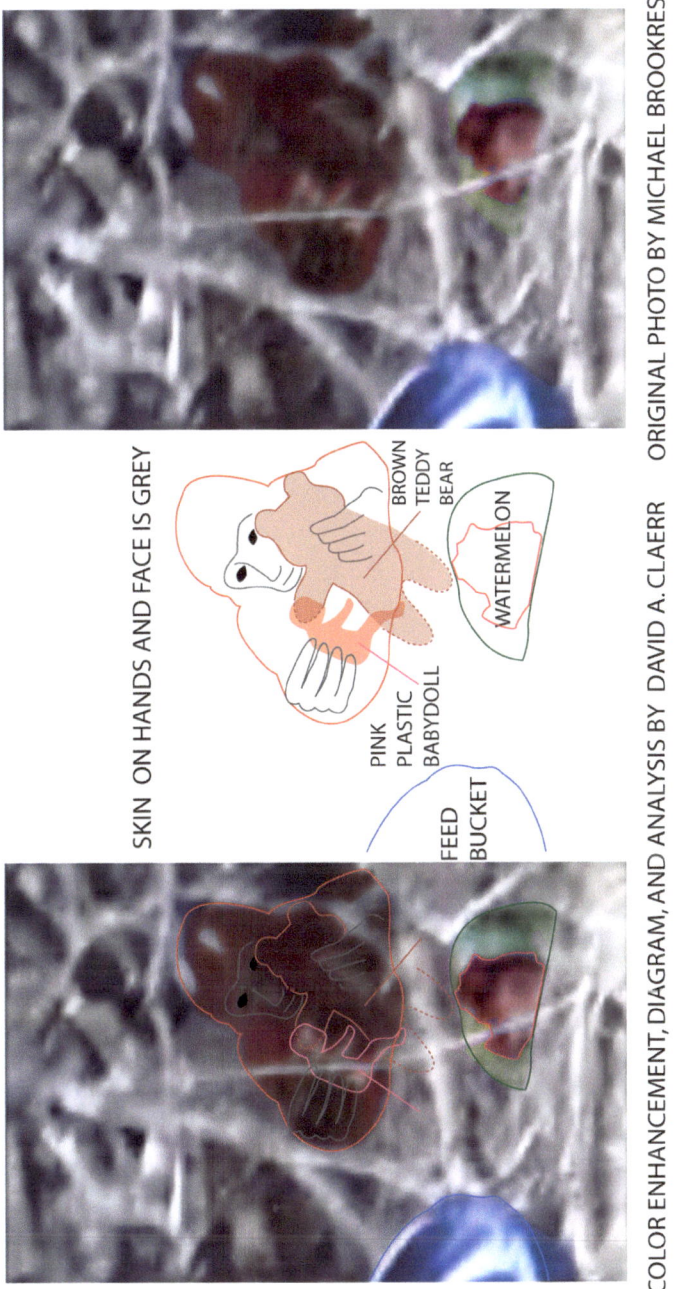

AN ENTERTAINING ADDENDUM

Often on exploratory expeditions, one has interactions with animals, both wild and domestic. Interactions can sometimes be both humorous and dangerous, as in this circumstance, which occurred about midway into our investigative search on the first day. When following the fence perimeters of the property on which we had permission to investigate, we could see a few horses peacefully grazing at pasture, inside a barb-wire enclosed corral. We passed by without disturbing them. We were making no attempt to sneak around the property surreptitiously, in fact, I was wearing a red sweatshirt for visibility- to avoid being a target for either poachers or legitimate hunters.

However, when returning along the fence line toward another part of the corral, there was a sudden thundering of hooves heading in our direction.

A stallion accompanied by a mid-sized colt burst through a gap in the fence. The terrain there was a mix of grassy patches interspersed with small clumps of cedar and post oak. The stallion challenged us with a trumpeting neigh, and headed straight for us, nostrils flaring and hooves pounding the turf. This was a serious situation. The horse was a full-grown and powerful specimen, on the attack. An animal that size can seriously injure or kill a person with a single punch of the fore-leg or a kick from the hindquarters.

In this encounter, to simply flee was out of the question, they would easily overtake us. No time to climb a tree, either. So partly from instinct and partly from experience in other such encounters, I decided to use a confrontational approach.

With a manner of gesture and tone highly reminiscent of the wizard Gandalf confronting the demon Balrog (in Lord of the Rings), I brandished my cedar staff and boldly commanded: "You Stop Right There- NOW!"

(OK, so that is not as cool as "YOU SHALL NOT PASS!", but it worked.) The stallion and colt were taken aback and hastily broke stride and came to a halt, which is not easy to do when at full gallop. The stallion looked at me wide-eyed, uncertain what to do. The colt fidgeted nervously.

I continued speaking in a more civil but insistent tone: "We are just passing through and mean you no harm. We will be on our way now." I turned away, glancing back over my shoulder after a few steps. The stallion began to stomp and snort, and seemed to be getting his courage up again. He made a small jump forward, getting ready to charge. I then wheeled around, and with a stern, loud voice proclaimed; "I TOLD YOU TO STOP RIGHT THERE, AND I MEANT IT !!

Once again, the horse was flummoxed by this gambit. He stopped, looking more confused than ever. (Keep in mind, this was a domestic animal, used to responding to commands.) So then I took a more considerate and conciliatory tone, as if I were a horse-whisperer:

"We are leaving now. You are a fine horse, a beautiful horse, and your colt is adorable. We mean you no harm. Nice horse, wonderful horse." (I almost laughed at myself there...) This also had a desirable effect, though. He made a brief snort, shook his heard several times, and then turned back, heading into the pasture corral, the colt following. We continued on our way, albeit, a bit shaken.

Notes on the following expedition:

In November of 2014, I made a follow-up visit to the same site. A series of tracks had been discovered by my host. The footprints in the tracks were exceptionally well- preserved, with the intricate detail of the dermal ridges. I had the opportunity to photograph the foot-prints as well as some of the hand-prints. I also made a plaster casting of the best foot-print. The dermal ridges (though in small patches) came out virtually as exactly they were in the print. Also visible are stress creases from the ball of the foot and down toward the arch.

The subject, which was undoubtedly a juvenile, had been foraging in the shallows for stranded aquatic creatures which might include fish, crawfish, freshwater prawns, frogs, clams and insect larvae, like dragonfly nymphs.

The hand prints and other traces in the sandy soil looked like the fingers were striking the silt and sand when our subject was grabbing for a fish or other prey. The foot-prints alternated direction and spacing, showing that the subject was striding forward, then turning and changing direction, then again, standing still. The tracks revealed that it left the area and returned a few times. This activity was rather plain evidence of a foraging/hunting behavior.

But the best evidence, I think is in the clear photos of the prints, because the area with the dermal details was a dark color, and the relief shows better because the highlights stand out more against the dark background. On the cast, the dermals are more difficult to see with the bright white on white of the plaster.

By having both sets of evidence, i.e. the photos and the cast, the reveal is stronger, in that the cast shows that the photos were not digitally altered.

(The photos of the prints and the casts from that foray are detailed, along with further analysis in a companion book, *Sasquatch In Texas: the Track Record*. The book also includes many other examples of cast handprints and footprints photographed and analyzed by the author. The book is available as a traditional hard-copy folio, and as an e-book for the Kindle Reader.)

www.ingramcontent.com/pod-product-compliance
Lightning Source LLC
Chambersburg PA
CBHW041610180526
45159CB00002BC/797